The Social Ecology of the Scientific Age

Miguel A. Sanchez-Rey

Table of Contents

Racism Amongst the Royals

5-7

An Integrated Pattern of Behavior Consistent with an Unprecedented Cult of the Scientific Age

9-11

The Terraformic Process

13-14

Innaturalism

16-17

Social Ecology in the Scientific Age

19-21

The Internationalist Model of The Second Task

23-25

Social Ecology and the Planetary Super-State

27-31

Racism Amongst the Royals

[Author: Miguel A. Sanchez-Rey]

The selfish gene assumes that animals are motivated to pursue selfish ends for reproduction and survival. That is selfish interests are the motivational paradigm of Darwinian evolution. But yet the early period of the Scientific Age has shown severe racism and bigotry in the scientific process, especially the Royal Society of London for Improving Natural Knowledge. Carried away by racial supremacy in ways that were never seen before since the rise of Nazi, Germany before the eve of the Second World War. Racial supremacy which stipulates favoritism over a racial/minority over another base on class, wealth, education, privilege and/or background. Consider to be the most violent form of racism.

A vestige of the classical era, in which due to racial supremacy, certain races/minorities are single out and excluded from consideration or fellowship. That said the royal family of the United Kingdom has been heavily involve in a formalism of racial supremacy since the Second World War in which King George VI symphonized with Nazi, Germany's policies of anti-Semitism. But it was only until Germany became an existential threat to the British empire that war was declared against the reigning German government.

For royalty assumes privilege as long as they oblige by the policies of the state, for they are heads of state. But they no bounds. By not respecting those bounds the Royal Society of London has become a cult institution in the form of a debating body.

Permanent damage has been done and the United Kingdom has falling ill that has cause a rising world-wide movement in favor of racial supremacy that has tarnish the Commonwealth of Nations and has setback unity amongst the developing countries that has giving into ultra-nationalist fervor in order to cope with the extreme racism in the develop countries, especially in Western Europe and the United States, spark by racial supremacy amongst the royals.

That is the Royal Society of London will be long forgotten after the Scientific Age. And their contributions in the scientific process will become ever more suspect to the Advance Physicists that will look upon the early 21st century as a critical period in human history where knowledge becomes not a matter of extreme selfish interests but of rational existentialism. In which bigotry and racism is of little tolerance and the Royal Society of London, and the learn institutions, must pick up the pieces of a fraudulent and shattered institution of the classical era.

An Integrated Pattern of Behavior Consistent with an Unprecedented Cult of the

Scientific Age

[Author: Miguel A. Sanchez-Rey]

The sociopathic norm sets the norm. A cult that behaves like a Cloud-Atlas mosaic that is very much a crime ring. They are academic and humanitarian cult-figures involve in both academic and financial fraud. Frighteningly messianic when first spotted they present themselves as loving creatures, with false claims of altruism and benevolence, but capable of horrid acts of hatred and malice. Completely oblivious to their actions they resort to different fraudulent means to further their fame and fortune. Controlling each other by imposing brutal tactics meant to keep a group from leaving or exposing the cult. Nevertheless, a scattered cult, they can seem oblivious and unaware of each other but share an interrelated existence that ends in murder and genocide. By which their followers are complacent in the actions of the mosaic of the norm.

They are a cult-phenomenon of flaw-decision makers that went world-wide at the dawn of the Windows XP revolution. Which has led to risky decision-making with the decline of the religious state to the fruition of a pathological mass-movement – with a rising favoritism toward racial supremacy, at a planetary scale. Knowing no bounds, they pose an existential threat to the natural and social order since they will tear apart the Scientific Age to justify their actions and existence.

The only avenue, to prevent cry havoc amongst their followers, that may lead them to mass suicide when the mosaic is expose, is to break them apart and prosecute the sociopathic norm to regain natural and social harmony in the Scientific Age.

The Terraformic Process

[Author: Miguel A. Sanchez-Rey]

The biosphere and much of the physical world is a terraformic machine. A machine that in its design means that achieving the terraformic process from an artificial standpoint is much easier than thought. That can be achieved in ways that are less costly but more efficient. PHPR [The Physicalist Program] aims to achieve a terraformic process in the form of a technological task understood as The First Task. A task that will utilize The Grand Unification Scheme to manipulate matter at the Ad [superstring] level.

Known theoretical options to achieve such a task has pondered processes such as nanotechnology and the melting of the southern and northern polar icecaps at the planet Mars. But such processes are arduous and ineloquent to human survival and longevity. The goal is to achieve a terraformic process that can be sustainable indefinitely through Incalculability and can be harness through technological means which allows for near-perfect manipulation of quantum matter. A process that takes only days to complete in its initial formalism and within decades a matter of minutes. That gives way to accelerating advances in engineering and technology that will spur habitability in harsher environments unlike the crude biosphere of planet Earth within the solar system and beyond.

Innaturalism

Miguel A. Sanchez-Rey

The natural world is a biological machine that is self-sustainable, finite, and simple. A simple machine that can be regarded as a biotic organism that adheres to molecular principles consistent with Darwinian evolution and epigenetics. That is principles are set in place that organizes the natural world into categorifications that give way to differential anatomical features. Anatomical features that are the byproduct of early adaptional mechanisms. At that end biological systems are complex driven. Which are emergent in its brevity and dynamical in its logical intuition.

At a certain point innaturalism is the antithetical of natural Platonism. Which artificial processes that are antagonistic to naturalism lead to innatural processes that surpass natural law but in which its breakdown becomes more and more ominous. That said natural Platonism stipulates the selfish aspect of biological evolution but also in which existentialism posits the nihilism of naturalism. Nihilism that stipulates the self-sustainability, finite, and simple processes of a biological machine that over a given period of time becomes innatural to the biological world. Which surpasses itself but in which its superiority leads only to an accelerating breakdown of the natural processes of complex and emergent biological systems.

Social Ecology in the Scientific Age

[Author: Miguel A. Sanchez-Rey]

As mineral resources reach its peak the Earth's biosphere arrived at a breaking point much earlier than expected. A breaking point in which the Earth's environment, including biodiversity and natural habitats, has begun to deteriorate at an accelerated rate due to the sudden innaturalism of homo-sapient industrialization, deforestation and population growth.

Climate change only exacerbates the Earth's ecological stress-levels that at that point, with the climax of mineral resources, the planetary system begins to decline. In which its downturn is due to persist until one reaches a dead-planet if no action is taking to reverse and mitigate those stress-levels.

The only possible action is to fully implement PHPR [The Physicalist Program] as the resolution to a foreseeable catastrophic scenario in the Scientific Age in the form of a task. The First Task is a 100 Year Task. A technological task which aims to resolve mineral depletion by using ITER [International Thermonuclear Experimental Reactor] as a 40-year window of opportunity that gives way to gradual environmental recovery and 60 years for both careful and full completion of The Grand Unification Scheme. In which there will be a slow global decline when ITER goes off the manufacturing line.

In that manner the Earth, in itself, is a terraformic machine. A terraformic machine that is self-sustainable but fragile and crude in its design. For such reasons achieving the terraformic process is easier than thought and much easier than one, two, and three. The resolution to the crisis of mineral depletion, which has cause the Earth to reach dissolution in its biosphere, is to pursue the eloquent and efficient endeavor to achieve a terraformic reaction through metaspace that will lead, by applying Incalculability, to a sustainable terraformic process at the Ad [superstring] level. Using The Grand Unification Scheme to manipulate and alter the Earth's biosphere that will give way to full environmental recovery that eventually results, by the mid-22nd century, the development of vast space-habitats in different strategically valuable areas within the Earth's solar system.

Doing so one cannot stress any further not to be carried away by Earth's biosphere or any other planetary biosphere outside the solar habitat, but to be wary that its much more efficient and eloquent to pursue the terraformic process by resorting to the vast construction of space-habitats. For the terraformic development of planetary biospheres is to be pursued only as a last resort and not as a means for long-term human survivability.

The Internationalist Model of The Second Task

[Author: Miguel A. Sanchez-Rey]

The internationalist model can be seen as the most likely resolution to The Second Task. The Second Task which has been establish as a resolution to an extraterrestrial conflict. In which the aim is to achieve co-existence with other intelligent life-forms outside the solar system and to begin the process of laying out habitable zones outside the solar system. Habitable zones that are scatter very much like island chains in the Pacific Ocean on planet Earth. Which will allow, with space-habitats, for human settlement and technological industrial development. Furthering the extraction of minerals and energy resources needed to sustain human life.

But in which one dares not interfere in the lively-hood of other intelligent creatures in the outer-habitable zones that if done will lead to a catastrophic conflict in which no way out is presentable to Earth's solar system and outer regions. That said the internationalist model appears as the most likely resolution.

The other avenue amongst astrobiologist and astronomers is either the development or acquiring of a star-map that will guide humanity through the solar system and throughout the Milky Way. But to even consider such an avenue requires the surveying of the Milky Way which will take, with feasible technological means, nearly 2 million years to complete and to acquire such a map will lead to genocide and/or a large-scale conflict. For a star map means the capacity to know everything there is to know about the outer habitable zones. For those reasons the star map is a dead enterprise and the internationalist model is the likely option in which resolution of The Second Task will lead to peaceful First Contact and eventually coexistence.

Social Ecology and the Planetary Super-State

[Author: Miguel A. Sanchez-Rey]

The planetary super-state of a Class 1 civilization is governed by a scientific council. A council made up of elected representatives from the workers' councils of the self-management of labor. They are elected officials that has authoritarian sovereignty over the work force of the planetary super-state. The council of the planetary scientific super-state takes whatever action is necessary to protect and uphold the democratic order. The democratic order that stipulates the democratic self-management of labor and the tranquility of the Scientific Age.

The planetary super-state as well has expanded further from their planetary biosphere. They occupy far from the outer rim vast space-habitats that encircle the solar system. In different strategically valuable areas a waypoint is situated that allows the planetary super-states citizens to move from Earth's inner solar system to the habitable zones. With millions that occupy Earth's space-habitats the human population of planet Earth has been declining since its peak at the end of the 21st century and Earth's ecology has since recovered from the environmental crisis of the late 20th-mid 21st century.

City-states are scattered in different regional locations on planet Earth while much of planet Earth are deserted regions of tropical and temperate forests whose population have chosen to forsake the technological advances of the Scientific Age and have remain content with the natural biosphere of planet Earth. While the city-states are technological advance regions with vibrant communities, buildings and sky-scrapers. They thrive as centers of commerce and trade with a wildly-strong attachment to their mutual co-existence and reciprocity.

The weather of the planetary super-state has been harness to suit the needs of the planet and with nuclear fusion the unlimited use of clean-energy has allowed humanity to live comfortably and care-free from the turbulent weather patterns of the 21st century by utilizing the terraformic process. Little is remembered about the famines that was spark by the agricultural decay cause by droughts due to climate change and mineral depletion or the rising ocean temperatures that almost subsume the major cities with growing flood-waters of the mid-21st century.

Human beings live independent lives with an attachment to the environment, to the sciences, and the democratic self-management of labor that has seen the dismantling of nation-states into regional domains that is the federation of industries of labor and workers' councils that are obligated to preserve the rights of future generations, and to protect policies that favor environmental and democratic equity.

With a command economy, that furthers New Keynesianism policies of secular stagnation and the development of perpetual economics, population growth has remained static in the planetary super-state but with the construction of space-habitats and the growing settlement of the habitable zones much of humanity on planet Earth has remain free from the travesties of crowded-life that plagued the early 21st century. Applying architecture to pacify the general population and using bioengineering to make city-life less polluted. Where environmental regulation forbids the misuse of microbes that may cause havoc to the environment or to preserve city-parks so as to keep the general population in line with the laws of the city-states -- and that, no violent uprising will arise that may directly threaten the democratic order.

But even then city-life is not an easy endeavor, for the democratic work-force must contribute to economic growth but remain in many respects participants in the scientific process of the environmental sciences. Cities are structured to meet the monetary needs of the planetary super-state so as to preserve tranquility, and so that the standard of living is maximize and that humanity may achieve indefinite life-expectancy in the bio-sphere of planet Earth.

Though neither tricked by inefficacy and ineloquence of the biosphere much of the population of planet Earth remain vigilant of the turmoil of solar weather and unforeseen cosmic catastrophes that may spell doom for the citizens of the planetary super-state. For those reasons, much of the population of the biosphere find life in the habitable zones and the space-habitats to be the ideal and that in the near-distant future planet Earth will be seen as a deserted eco-system while a new settlement for the human race as a whole lies out far into the distance of the galaxy in which over time humanity will become amnesiac to their past history on planet Earth presiding as an incalculable civilization that far surpasses itself in its wildly strong temperament and resilience. But no further.

While weeds and plants overtake the city-states and oceans causes bridges and buildings to crumble as the still weather echoes the windy sirens of a desolate planet that has since forgotten itself. There humanity will be far from its reach but even then far different from the human species that was born of the biological processes of evolution.

www.ingramcontent.com/pod-product-compliance
Lightning Source LLC
Chambersburg PA
CBHW062208220526
45470CB00009B/2972